Reflections of Life

from the heart of a country girl

Verna Dycke

authorHOUSE®

AuthorHouse™
1663 Liberty Drive, Suite 200
Bloomington, IN 47403
www.authorhouse.com
Phone: 1-800-839-8640

First published by AuthorHouse 6/12/2007

ISBN: 978-1-4343-0243-4 (sc)

Printed in the United States of America
Bloomington, Indiana

This book is printed on acid-free paper.

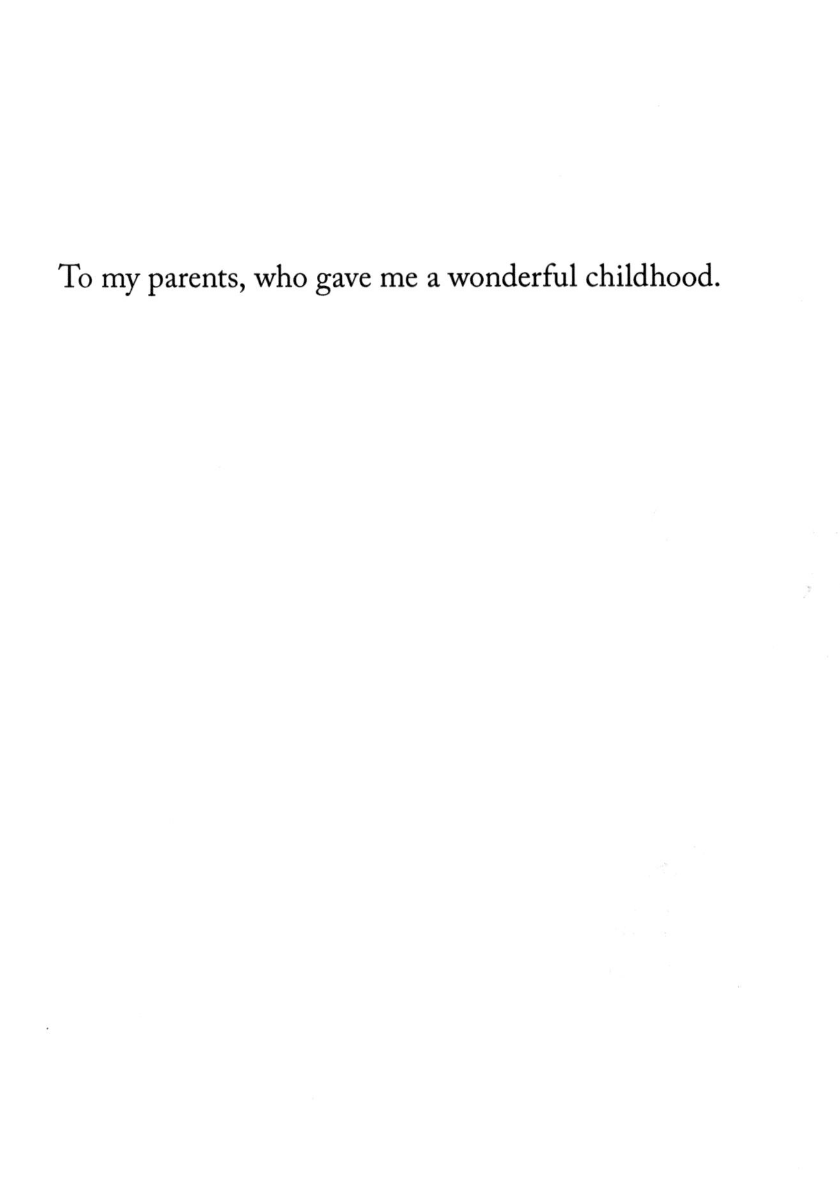

To my parents, who gave me a wonderful childhood.

Table of Contents

Neighbours

PETS

I Live Here (Palindrome)

I live here
Where lions roar, eagles soar,
Earth meets sky, grown men cry.
I live here
Where all is love, heaven above.
Bye and bye angels fly.
I live here
Where dreams come true, me with you,
Never die. Stars on high.
I live
Here
Live I
High on stars. Die never,
You with me, true. Come dreams, where
Here live I.
Fly angels, bye and bye,
Above heaven. Love is all, where
Here live I.
Cry, men grown. Sky meets earth.
Soar, eagles. Roar, lions, where
Here live I.

To Paint
(Poetry In Motion Contest Poem)

This spirit came
Dressed with living light,
Free thought,
Laughter,
Embraced understanding,
Willingness to love
Swelling in its unconscious,
Stirring a delicious desire
To paint
What only I can see;
Those jewels of silky illumination
That glittered
Into the deepening twilight
Like shades of evaporation;
Cool, blue breath,
Hot, white sand,
Soft-buttered sun,
Golden glow,
Bronze, smoky haze,
Ripening-plum pollen,
Moving slow motion
Across the dreams
Of my life.

"Springborn,"

I arrived in Spring ,
Independent, energetic,
intelligent, competitive,
I like family, friends,
children, animals.
I dislike manipulation,
sloppiness in spelling, writing.
I find it difficult, believing
I'm aging this quickly.
I will remain optimistic,
Laughing, giggling,
Enjoying birdsongs,
Crickets chirping,
Butterflies fluttering,
Children playing,
Until I die.

A Challenge Poem - Write a poem about yourself and every word must contain the letter "I"

My Autobiography (Abcderian)

Animal-lover, artistic, aging (gracefully, I hope!)
Brown hair, blue eyes, brat (growing up), bookkeeper
Christian, caring, compassionate
Daring, different, dependable
Eager, enthusiastic, energetic (used to be, now it's used up fast)
Foster Mom (for the past 15 years, to over 100 children)
Generous
Helpful, happy, humourous (not everyone gets my sense of humour)
Intelligent, interested (in many things)
Joyful
Kind, knowledgeable
Little darling (meaning of my middle name, Darlene)
Musical, mathematician (my favourite subject in school)
Nature-lover, naughty (as a child)
Outspoken (sometimes), overweight (trying to change that!)
Poet, perfectionist
Quiet, quaint, quarrelsome (trying to conquer that!)
Rebel (against the things I feel are wrong)
Spring born (May 1, 1949), single, short (5' 1")
Thoughtful, thankful
Understanding, unique
Verna (my first name)
Wholesome (my friend says so)
X-ray vision (my kids think I have it. Don't tell them otherwise!)
Young-at-heart
Zealous (for what I believe in)

Fostering (Villanelle)

The kids I choose to raise are not my own
They come to me with problems great and small
They need a home where real love can be shown.

Sometimes abuse, neglect is all they've known.
Just thinking of the horror casts a pall.
The kids I choose to raise are not my own

Sometimes they're cute and cuddly, all alone,
Or maybe a big bunch who love to brawl.
They need a home where real love can be shown.

I try to make my place a comfort zone.
Where they can learn to play with truck or doll.
The kids I choose to raise are not my own

At times I pull my hair and moan and groan,
And wonder, "Am I any help at all?"
They need a home where real love can be shown.

I pray that they will grow, the seeds I've sown
In hearts so often guarded by a wall.
The kids I choose to raise are not my own
They need a home where real love can be shown.

My Children Dear (French Rondolet)

To my foster children.......

My children dear,
I grew to love you as my own.
My children dear,
I had you for just three short years
And then you left me all alone,
But now I see you fully grown,
My children dear.

But you are His.
I came to think of you as mine,
But you are His.
An error of the heart, it is.
To God's will I did not resign.
For many months my soul did pine,
But you are His.

You loved Him then.
Though now you've gone so far astray,
You loved Him then.
I know He'll keep you to the end.
My job is just to trust and pray,
That you'll return to Him one day.
You loved Him then.

Crazy Dreams

The Well

A child, I fell into a well
No water therein, which was swell,
But how to get out? Couldn't tell,
Not deep, but steep! I tried to yell.

No answer came. I made a pledge,
"I'll get MYSELF out!" I did hedge.
I found a ladder, and a ledge,
And climbed up swiftly to the edge.

The School Bus

The school bus careened down the road,
I'm not sure how full was her load,
For I sat on top,
And hoped she would stop,
But never a smidgen she slowed!

The corner was coming up fast,
And I thought that I'd never last.
She had super powers,
Came to rest in our flowers,
And I stayed intact! What a blast!

The Bull

The bull was raging mad
But all the strength I had
Was not enough to make my legs propel
My heart beat fast in fear
As quickly he came near
And when he was upon me, down I fell.

I lay there, still as death.
I could not take a breath,
'Til suddenly I knew what he had done!
He'd jumped right over top,
Kept running, didn't stop,
And left me safe and sound! He'd had his fun!

Allergies (Essence Suite)

When gentle breezes blow
I'm going to sneeze, I know.

My allergies will flare
And coughs and wheezes, tear.

The dandelion bright?
Can't stand the lovely sight!

Its round, white head appears
As summer stead'ly nears.

To Pharmacy I'll lope,
Disarm it with some dope!

An antihistamine
Will manage well this scene!

Oh, sweet drug of relief.
I'm freed beyond belief!

“Goodbye” (Cameo)

Why are
All my friends leaving?
Seems that when I start to feel close
To someone
They pack up and move far away.
Will I ever get used to
“Goodbye“?

Skateboarding (Trine)

My son is into skateboarding
He wears the clothes, collects the bling.
The skateboard park his second home,
There's nowhere else he wants to roam.
He spends his money every week
To make his board look most unique.
"God, keep his arm out of a sling,
And bring him always safely home.
Your favour, Lord, is what I seek."

It's Back To School Again

It's back to school again.
My books are in the bag
It's back to school again.
With pencil and with pen
I head for the unknown.
My books are in the bag
Who will my teacher be,
My classroom and my desk?
I head for the unknown.
Excitement in the air.
Can't wait to see my friends,
My classroom and my desk.
The air is crisp and cool
My step is quick and sure.
Can't wait to see my friends.
The old routine returns.
It's back to school again.
My step is quick and sure.
It's back to school again.

Neighbours (Trine)

For over twenty years my life
Has been most free of stress and strife,
New tenants now reside next door,
And peace and quiet reigns no more.
There's yelling, cursing, music loud
And visits from a motley crowd.
I fear one day he'll use a knife,
Or gun or club, and go to war,
With all the hatred he's allowed

"Oh, Father God, You see this scene,
And know where this young man has been.
The sorrow, pain, abuse he's felt.
The gross unfairness he's been dealt.
I ask You, shine Your love on him.
Lord, fill him up right to the brim.
Remove each thought of being mean,
And cause that hardened heart to melt,
That he may know, life's not that grim."

"His girlfriend, Lord, I bring to You.
I've known her since she was just two.
Why does she let him put her down,
Defending him without a frown?
Please show her, Lord, how true love feels,
And that, all her past pain it heals.
Her broken heart, create anew,
And then Your princess, proudly crown,
As at Your feet she humbly kneels."

"Amen"

Scrapbooking (Rondeau)

My latest hobby so consumes my time
I work from dawn until the midnight chime
My table's piled with paper, scissors, glue,
And where to find that picture? Not a clue!
I've left the lid off! Glue has turned to slime!

I've gathered decorations, purple, lime,
With photos from the days of "Five and Dime,"
And brightly coloured pages, green and blue
This scrapbook phase, it so consumes my time

Some poetry I'll add, with metered rhyme,
I want this gift to be just perfect, prime,
So ponder each decision, fret and stew.
Each page must have the best design and hue.
Then wearily to bed I slowly climb.
This Christmas project so consumes my time

Growing Old (Internal and External Rhyme)

I'm growing old. The weather's cold and hard to tolerate.
I must confide, I stay inside, just read my book and wait.
With Spring's return again I yearn for sun and balmy weather
I'll go outside, no more denied. I'll throw off Winter's tether.

I'll pull the weeds, I'll plant some seeds, I'll take a little walk.
I'll meet a neighbour on the way, we'll stop a while to talk.
My coffee cup, I'll fill it up and sip it 'neath a tree.
The sun will shine. 'Twill be so fine to suddenly be free.

Housework (Ends Edge)

It never is much fun for me
Or we would have a house so clean
No jeans be lying on the floor
Before or after bath or sleep.
Oh, keep your comments to yourself,
Your shelf has got a speck of dust!
I must bring your attention to
The crew that works to clean your house!
A mouse would never find a crumb,
The hum of vaccum always heard!
Absurd! I like my messy home!
We roam and play inside and out,
We shout and roll on grass and hay.
Away, house cleaning chores, just go!
I know you'll still be there for me
To see to, come the rain or snow,
Tomorrow!

Procrastination (Canzone)

My house is topsy-turvy,
The dishes piling up, the floors need sweeping.
I sit and read my book while tea is steeping.
But soon I'll hurry, scurry,
And tackle this big mess that's slowly heaping,
But I won't worry.

My strength, it often fails me
To work the whole day through, for endless hours.
I sit outside enjoying birds and flowers.
I watch my favourite movie,
Or take a nap, then play a game of Yahtzee,
Just take it easy.

I used to get to stewing
If everything was not in place and spotless,
But now I go about my day quite thoughtless.
I do what most needs doing,
And don't let what's undone be my undoing.
I'm done with stewing.

(I have an auto-immune disorder, and chronic fatigue is one of the effects I have to deal with. The only way I can get through the day is to pace myself, and stop "doing" when I start to feel tired.)

Our Family Life

We didn't go to movies, we didn't have T.V.
There was no indoor plumbing or electricity.
We didn't join the other kids in after-school events,
Because we had a life at home, complete in every sense.

In Springtime there was planting, the garden and the field,
And also fertilizing, so better crops they'd yield.
New calves were born, and kittens, while fuzzy chickens hatched.
We cared for them, adored them and quickly grew attached.

Then Summer came, with weeding and berry picking too,
And canning and preserving, all things that we must do.
Haying time was busy, we worked from morn' to night,
Then vegetables were gathered, in a cellar void of light.

When winter came with cold and snow, we had a chance to rest
We'd dress up warm and go outside to slide or ski with zest.
We'd tramp through snow up to our waists to find a Christmas tree,
And search for tracks of squirrel, hear chirps of chickadee.

My Dad would get up early, make a fire in the grate,
And then he'd put the coffee on. The rest of us would wait
Until the house was cozy, and Mom had breakfast on.
Then we'd all eat together, and off to school be gone.

We were a close-knit family, did everything as one
When all the chores were finished, we had a lot of fun.
Our Dad would read aloud to us, we'd play a game or two.
While Mom would knit us socks or mitts, we'd wind the yarn, brand new.

We'd listen to the radio, with ear pressed right up close,
As Billy Graham preached and Johnny Cash his love would boast.
At bedtime we were tired, and as soon as we had prayed
We'd fall asleep beneath the woolen quilts that Mom had made.

We didn't go to movies, we didn't have T.V.
There was no indoor plumbing or electricity.
We didn't join the other kids in after-school events,
Because we had a life at home, complete in every sense.

My Parents (Triple Rictameter)

My Dad:
Lover of life,
Watching over God's Earth;
Flower gardens, fields and forests.
Loving children, nine in total. Playing,
Racing to the house after chores.
Feeding cows and horses
Before eating
Our meal.

My Mom:
Loved her fam'ly.
Cooking, cleaning, sewing
New clothes from old. Baking fresh bread;
Wonderful gardener, beautiful flowers;
Berry picking, canning surplus;
Knitting socks and mittens;
Playing card games;
Reading.

Parents:
Making a home;
Teaching us right from wrong.
Having coffee, helping neighbours.
Working together, talking things over;
Not one fight do I remember.
Taking time for picnics,
Tramping old trails,
Mem'ries.

Isaac Dycke
(Jan 12, 1897 - Jan 23, 1965)

My father,
A man of the soil,
Cared for all of creation
With great devotion.
Flowers and trees,
Birds and bees,
Streams and seas
Were to be treasured
And protected,
For it was from them
That we received
Our daily bread.

I am honoured to be
My father's daughter.

My Mother

My mother,
Cooks, cleans, cares for her children,
Gardens with
Great pride in the produce grown
And lovely flowers.

My mother and I,
Learning together,
Grades one, two and three.
Studying at home,
Reading, writing, 'rithmetic,
Too far to attend the nearest school.

My mother
Finds joy in reading, painting,
Visiting
Neighbours who stop for coffee,
And her tasty treats.

My mother and I
Manage the farm, with
Sister, brother and
Good, helpful neighbours,
After Dad is gone. It is
A time of much adjustment and change.

My mother,
Makes sure we finish Grade twelve,
So that we
May have opportunities
She has never had.

My mother and I
At my High School grad
She is very proud.
I am the first one
In my family. "Now go
And do something worthwhile with your life."

My Mother
Always there for me, making
Sure that I
Have all that I need, even
After I am grown.

My Mother and I,
Friends right to the end,
Even though at times
We do not agree.
She lets me be me, and loves
Me always, unconditionally.

Anna (Klassen) Dycke (March 1, 1913 - April 13, 1992)

Our Mothers come in every shape and size.
Now, some are young, but others old and wise.
Today we take the time to recognize
Just what they mean to us. Oh, how time flies!

My Mother, rest her soul, was always kind,
To many of our childish errors, blind.
She loved us, and the time she'd always find
To play a game or hear what's on our mind.

She never gave us up when we were bad,
Our childhood was the greatest time we had.
Our Mom was always there for us and Dad.
She even liked our music, just a tad!

When we grew up and went on our own way,
She always called to ask, "How was your day?"
She'd help us out if hardships came our way,
And lend an ear so we could have our say.

Though Mom's been gone now over fourteen years,
And time has stopped the flow of sorrow's tears,
Remembering her love calms all my fears.
"Dear Mom, you are the best." I pray she hears.

My Childhood Home (Petrarchan Sonnet)

Our country home, it means so much to me.
The old log house, "So cozy," they'd exclaim.
While growing up, we felt so unashamed,
And ran and skipped so thoughtless and carefree.
The neighbour kids all welcome, "Come and see."
Our yard was large enough for softball games,
Or relay races. "Tag!" We called out names.
And played away the days, with joy and glee.

The barnyard was a special place for me,
With cows and baby calves, so very fine,
Or riding on old Red. Though very slow,
He took me many places, roaming free
The woods, another favourite haunt of mine.
My childhood mem'ries, still, I love them so.

Family Limericks

The Hunters

The Uncles would come to hunt moose,
With stories of escapades loose.
We listened in awe
As they'd break every law,
And then from the scene they'd vamoose!

Checkers

I loved to play checkers with Dad
And a challenge from Roy made me glad.
The adults sat 'round.
There wasn't a sound,
As I easily beat the poor lad!

(I realize, now that I am grown up, that this may not be exactly how the game went. Roy was 2 years older than I, and he may have deliberately allowed me to win.)

Superstitions

We'd laugh with great joy and delight,
And Mom, being true Mennonite,
Would worriedly say.
"Don't get carried away,
Or bad news will come here tonight."

"Don't step on a spider. It'll rain!
Walk under a ladder, feel pain!
Set thirteen at table?
Our Mom wasn't able,
For soon one would surely be slain.

(These superstitions of Mom's used to really drive me crazy, and I would deliberately do the things she was so afraid would bring bad luck, just to prove them false.)

To My Aunt Justine (Dec 30, 1920 - May 4, 2005)

So tiny and frail, and yet so strong,
You beat all the odds, and lived so long.
Disabled from birth, It was hard to walk,
But still, you worked as a waitress and cook.

You saved up your money, and off you went,
To Rochester, Minn. with one intent,
To go through the surgery and mend your hip,
And, all said and done, it was worth the trip.

Returning, you married and bore three kids,
Raising them well. Much mending you did.
Your garden was one of the nicest around.
Its flowers were always the talk of the town.

You sang in the choir, taught Sunday School, too.
The Pioneer Girls all looked up to you,
And then, when the grandchildren started to come,
You welcomed each one of them into your home.

You were such an example, right to the end.
You always found time to be a good friend.
Whatever the need be, a joke or a prayer,
You always had just the right words to share.

And now that you're gone, we'll miss your bright smile,
But know that we'll see you again, in a while.
Until then, our memories will have to suffice.
Just knowing you, Auntie, has been really nice.

My Favourite Uncle

No one could make a person laugh
Like my favourite Uncle, Don.
His witty sayings filled the house.
I miss him, now he's gone.

"De-BO-rah, De-BO-rah
DON'T spit on de FLOOR-ah
USE de cuspi-DOR-ah
DAT is what it's FOR-ah."
Or
"When I say, 'Jump,'
The kids ask, 'How high?'
On the way up."
(Never happened!)
Or
He would come over,
Look down on one of us short people,
And say, "Are you standing in a hole?"
Or, "How's the weather down there?"
(He was more than six feet tall)
Got us going every time.

No one could make a person laugh
Like my favourite Uncle, Don.
His witty sayings filled the house.
I miss him, now he's gone.

Uncle Herman, Unknown Soldier (Englyn Penfyr)

About my Uncle Herman I know nought.
He fought in Europe, to die.
An Unknown Soldier, to lie.

No funeral, not family nor friend
Life's end, to mourn on bent knee.
To pray, and set his soul free.

He was a child, a lad of just sixteen.
So keen was his wanderlust,
That, lie of his age, he must.

There is a portrait hanging on the wall.
So tall and handsome, but gone,
His loving mem'ry lives on.

Back Home on the Farm (Epulaeryu)

Roast chicken with sage dressing,
Carrots, peas and greens
Gathered fresh from the garden.
Freshly baked bread with
Sweet, home-churned butter.
Berries and
Cream!

Picnic Time (Epulaeryu)

Peanut butter sandwiches,
Cheese and tuna too,
Carrot sticks and celery,
Chilled watermelon,
Lemonade to drink,
Summer day's
Fun!

Home

Home is the place where you feel all at ease;
You can say what you think, you can do what you please.
No matter how far you may travel or roam,
There's something inside you that's calling you home.

It's there that your memories are happy and gay;
It's there you're made welcome anytime, night or day;
That when you grow tired of wandering alone,
You can leave all your cares and say, "I'm coming home."

You'll make many friends, and learn many things;
A lot of experience, travelling brings.
But even if all of your lifetime you roam,
There's still just one place in the world that's called home.

Neighbours

Daisy, Daisy (Song)

Daisy, Daisy, where will you go today?
Walk for miles and visit along the way.
You don't have a horse to ride on,
Nor even a sled to slide on,
But someday soon
We'll sing the tune
Of a three-wheeler built for you.

Daisy, Daisy, roaming the countryside,
Hill and valley, visiting far and wide.
The neighbours all hear you coming,
Your little motor humming.
You do look sweet
Upon the seat
Of that three-wheeler built for you.

PETS

Pup (limerick)

A small, feisty Terrier was Pup.
Together we played and grew up.
He loved fetching sticks
And chasing the chicks,
And sometimes on one he would sup.

Now Pup's proper name, it was Sport.
His tail had been cropped very short.
He'd hunt in the hills,
Get a face full of quills,
Removal brought nasty retort.

His life, it was long and fulfilling.
A watchdog superb, always willing.
He lived fifteen years,
Then killed by his peers,
His violent end was quite chilling.

Ollie (limerick)

Ollie, a mutt black and white
With long fur to block his eyesight,
Would pester and tease,
'Til Pup he'd displease,
And then would erupt a great fight.

The racket was scary, for sure,
To death they both would have endured.
A bucket of water,
The colder the better,
Discovered to be the lone cure.

Dagwood (limerick)

Our Dagwood, Chihuahua and Terrier,
Would run, jump, and play, ever merrier,
But when teeth would bare
You'd better beware,
And avoid his invisible barrier.

Rusty (limerick)

Our Lab-Collie pup we named Rusty.
He grew to be gentle and trusty.
He herded the cattle
With never a battle,
'Til one day we lost our dear Rusty.

The cows would all run here and there,
The way to go home, unaware.
We searched for that hound,
Until he was found,
And back herding cows with great care.

Tinker (limerick)

My Tinker, part Pom, was a stray
Who moved into our home to stay.
With long golden tresses,
She loved soft caresses;
A beauty, in her own haughty way.

Three puppies she bore in a while.
Their antics made all of us smile.
But, "Don't you come near,
Or you'll learn to fear!"
She could be quite nasty when riled.

My brother, when bringing in wood,
Would try to avoid, if he could,
A nip to his heel,
The pain very real!
Proud mother protected her brood.

Sleeping in the Hayloft

Sleeping in the hayloft,
Kittens everywhere
Batting at my eyelids,
Tangled in my hair.

Underneath the covers
Chewing on my toes;
Catch a tiny wink. One's
Nibbling at my nose!

Three a.m. and counting,
Still no sleep in sight.
Bundle up my blankets.
Back in bed. Sleep tight!

The Whispering Pine

I see the silver shadows in the soft moonlight.
I sway and swing my branches in the West Wind's height.
I shelter many squirrels from the Winter's bite.
I am the Whispering Pine.

I stand beneath the summer sun and clear, blue sky.
I see my friends turn yellow and prepare to die.
I snap and crackle noisily when Winter's nigh.
I am the Whispering Pine.

In Spring I whisper, "Welcome," to the nesting birds.
In showers I am shelter for the cattle herds.
To people who can hear I whisper wise, old words.
I am the Whispering Pine.

Skies are grey.
Friends have flown far away.
No ambition. "Please give me a shove."

Where is the joy? Where is the peace? Where is the love?
Not enough energy to seek help from above.
Look out the window, half in a daze.

Barrenness meets my gaze.
Dreary days.

A Midnight Walk

The first snowfall, as soft as eiderdown
Has blanketed the world, her pure, white winter gown.
The frost, like diamonds, sparkles on the grass;
Spreads lovely lace on every twig and branch.

The moon and stars, so far and yet so bright,
From cold and depthless skies, like lanterns, spread their light.
My path has not been marred, no tracks been made
Save those of tiny feet where rabbits played.

I stand in awe and marvel at God's love,
To send such wonders down to us from Heav'n above.
I know that nothing I may ever do
Could help repay one glimpse of such a view.

First Snowfall (Freestyle Poetry)

Feathery filaments
Of fluffy frost
Fall from above;
Soft flakes, flying,
Floating free.
Whee!

Whispers of white,
Wintery wool,
Whirl,
And twirl.
And swirl
On the wings
Of the wind,
Drifting,
Lifting,
Settling slowly down,
Covering the brown,
Cold ground
With a downy comforter.
'Til spring comes 'round
Again.

I Love You, O Wind

I love you, O Wind.
You speak to me of God.

I hear you rush through the trees,
Your voice strong and comforting
After the bitterness of Winter.
You whisper sweetly, lovingly,
Coaxing tender buds into view.

Sometimes you become harsh,
But you uproot only those weak,
Shallow-rooted or already dead.
Your anger causes the healthy
To send their roots deeper,
And become even stronger.

It is you who changes the weather.
One day you bring dark rain clouds,
But soon you blow them away again,
And the Earth is the more beautiful
Because of the cleansing rain.

Yes, I love you, O Wind.
Your every breath compares to my life,
And so you teach me
The ways of God.

Seasons (Double Diamonte)

Spring,
Fresh, new,
Budding, growing, blooming,
May, lilacs, beaches, baseball.
Swimming, camping, gardening,
Hot, humid,
Summer.

Autumn.
Windy, chilly.
Blowing, falling, dying
September leaves. Hockey, snowmen.
Shovelling, skating, sledding.
Cold, snowy,
Winter.

Gentle Breeze (10/28/06)

Gentle breeze, singing, winging
Hope throughout the land,
Spring, the Earth is growing, showing
Life on every hand.

Gentle breeze, warmly wafting,
Whispering my name.
Summer showers shimmer, glimmer,
Joining in the game.

Gentle breeze, blowing, flowing
Softly through the trees.
Autumn leaves, whirling, twirling,
Dancing with the breeze.

Gentle breeze, fresh and frosty
Nipping at my nose,
Winter branches sparkle, crackle,
All the world has froze.

Peaceful Morning (Englyn Penfyr)

Soft clouds float peacefully in pale blue sky
While bye and bye in the dale,
Red fox disturbs flock of quail.

Great flutter of wings as they quickly fly
And sly red fox tries new trick,
Moving leaf and dirt and stick.

Deer feast this morning on dew covered grass
While massive buck guards above,
Protecting his herd with love.

Redwing blackbirds sing early morning song,
Tall, strong bulrushes, adorn.
Tiny field mouse babes are born.

Eagle hovers, eyeing movement below,
But no small creature stirs. Scent
Of warm mist rises, night spent.

Spring (Clarity Pyramid)

SPRING
New life
Hope returns

Foals and lambs frolic,
Grass and leaves are greening,
Sun shines, gardeners rejoice.

“A time to plant and nurture life.”

The Leaves are Turning Green (Monchielle)

The leaves are turning green
The snow has gone away
It's time to laugh and sing
Of miracles of birth.
Oh, what a joyous ring!

The leaves are turning green
And growing every day
The sun is shining bright
Soon flowers will appear
And reach toward the light.

The leaves are turning green
For Spring at last is here
New Mothers guard their young
As carelessly they play.
And cheerful songs are sung.

The leaves are turning green
All gardeners' delight
The earth so soft and brown,
Is ready to be dug
And clothed in summer's gown.

The Pine Beetle (Limerick)

Our pine trees are fast turning brown,
And soon they must all be cut down.
Because of the beetles
They're losing their needles.
Whole forests will fall to the ground.

(This was a huge problem in the summer of 2006, and many trees were lost. Now, in 2007, many dead trees still need to be disposed of before forest fire season is here again.)

Forest Fire Season (Double Nonet)

Blazing across the dry countyside,
Consuming all that's in its path,
Robbing creatures of their homes,
Leaving mass destruction,
This huge inferno
Began, with just
One tiny,
Little
Spark.

But,
Next year,
Lush, new growth
Begins beneath
Blackened, charred debris.
The whole countyside is
Springing back to life, bringing
Birds and beasts to nest once again
'Mid pine trees sprouting from fire-popped cones.

The Thunderstorm (Haiku)

Dark storm clouds gather.
Large drops begin to rain down.
Suddenly it pours.

Jagged lines of light
Dart across the evening sky,
Bright arc overhead.

Loud crash of thunder
Splits the air, rattles and shakes
What can be shaken.

Finished just as fast,
Fires kindled all around.
Storm leaves destruction.

Autumn Days

I love to listen to the wind
As it whips through the willow,
Swish, swoosh, swish, swoosh,
And the chimes by my window
As they tinkle and clang in the breeze.

I love to walk
In the fresh, crisp, cold autumn air,
As its frosty breath nips my nose and cheeks,
Colouring them a bright, rosy red.

I love to watch the dying leaves
Take their final flight
As they gleefully glide,
Glowing red and gold,
And gather together on the cold ground,
In their departing glory.

I love to gaze up into the sky
As the geese gather in the familiar vee shape,
With honks to guide on their journey south
To winter grazing grounds.

I love Autumn,
With its own unique sights and sounds
That stimulate the senses like no other season.
It is the final burst of Earth's abundant life
Before the cold death of winter
Arrives.

Autumn Nears

As Autumn nears, Jack Frost appears
A cold wind rustles through the leaves
As Autumn nears, Jack Frost appears

I walk upon familiar ground
With rustling sound. Bright hues abound.
A cold wind rustles through the leaves

The geese are honking overhead.
While squirrels chatter, gath'ring cones
With rustling sound. Bright hues abound.

I sit and listen, breathe in deep,
The chilly air, its scent so sweet,
While squirrels chatter, gath'ring cones

My cheeks are rosy, and my nose
My fingers stiff and nearly froze,
The chilly air, its scent so sweet.

I turn around and head for home,
As Autumn nears, Jack Frost appears
My fingers stiff and nearly froze.
As Autumn nears, Jack Frost appears.

Quiet Time (Fibonnacci)

Peace of God (Fibonacci)
Dark of Night (Fibonacci)
Submit and Resist (Englyn Penfyr)
Who? (Rhyming Poetry)
Are You Satisfied?
A Morning Prayer
My Prayer
When I'm Alone
Your Will Be Done *
God's Love
Great Victory is Coming (Paradelle)
Treasures in Heaven
In His Image
Good Versus Evil
Let It Flow!
The Lord is Good (Rondeau)
Who? (Quinzaine)
The Prayer of Jabez (Pantoum)
Treadmill **
Make the Choice
God is Speaking (Autoinversion)
Believe (Rhyming Poetry)
Dry Bones (Rhyming Couplets)
Prayer for Governments (Deten Poetic Form)
Circle of Life (Double Eintou)
I Dreamed a Dream (Rhyming Poetry)
Love Sets Free
Love (Decimeter)
Joy (Decimeter)
Peace (Decimeter)
Patience (Decimeter)
Gentleness (Decimeter)
Goodness (Decimeter)
Meekness (Decimeter)
Faithfulness (Decimeter)
Self-Control (Decimeter)

Quiet Time (Fibonnacci)

My
Large
Mug of
Steaming hot
Coffee sits waiting,
Slowly cooling as I forget
I poured it, too busy with the everyday morning
Chores that seem so important, yet rob me of my more important
quiet time with God.

Peace of God (Fibonacci)

I
Walk
Along
The beach and
Hear the waves lapping
The shore, feel the cool sea breeze on
My face, see the gulls soaring and swooping overhead,
And I feel the peace of God that transcends all human
understanding, calming my fears.

Dark of Night (Fibonacci)

To
Gaze
Into
The dark of
Night, and see the stars
Twinkling brightly, lighting up the
Universe, is to know that God is real, with awesome
Power and majesty, and small as I feel in that moment, He also
cares for me.

Submit and Resist (Englyn Penfyr)

Submit to God and He will show His hand
Mighty and powerful still.
Cling to faith and know His will.
Resist and Satan will far away flee.
Never be tempted to spar.
Do not your faithfulness mar.

Who? (Rhyming Poetry)

Oh, you who don't believe in God,
Where do you go when times are hard?
When dreams are shattered, loved ones die,
Where does your consolation lie?

And if you don't believe in God,
When you feel weak, who lifts your load;
When everyone you thought a friend
Has left you, and life seems to end?

If, as you say, there is no God,
Who do you thank when you feel awed
By song or beauty, oh, so rare,
No common explanation there?

And if not God, then who's the one
Who makes the Universe to run
In perfect time, and who can shape
A mountain and a wee snowflake?

Are You Satisfied? (Free Verse)

You search and search
For freedom, peace and happiness;
You wander from city to city,
Town to town,
Into the country
And out of the country,
But nothing satisfies.
You buy houses, cars, clothes,
And many luxuries besides;
You save money in the bank,
You provide a good education
For your children,
And still nothing satisfies.
You live good lives,
Give money to charities,
Show kindness to others,
And are always willing to help
Someone in need.
You go to Church faithfully,
And everyone looks up to you,
But even then nothing satisfies.
You do not realize
That these outward appearances
Are hopelessly unable
To satisfy.
That true freedom, peace and happiness
Comes only from a heart
At peace with God.
Then,
Whatever the outward circumstance,
You will be satisfied.

A Morning Prayer

I'm weak and often doubtful,
And trouble then I find.
Oh, Lord, today please let me have
A strong and peaceful mind.

My thoughts are sometimes hateful,
My speech and actions too,
So Lord, today, a loving heart
Is what I ask of You.

At times I'm too self-centered;
Around me, fail to see
The beauty, so dear Lord, today,
Please give me eyes that see.

And then, too oft' I'm lazy.
The good that I could do
Is left undone; so Lord, to me
Give helping hands for You.

And if I fail to hear Your voice,
As I so often do,
My need is then for hearing ears.
These, too, I ask of you.

Now, Lord, I've asked for many things,
And so that You can do
What I desire, You ask but one.
I give myself to You.

My Prayer

"Give me a heart full of love, Lord.
Help me be happy today.
Let me not fuss or complain, Lord,
But praise You with each word I say.

"Guide me to selflessly give, Lord,
Just as You've given to me,
Happiness, love and forgiveness;
The things that help set a soul free.

"For, Lord, with my heart overflowing,
Then only will others receive
The love that is poured out upon them,
And then they'll begin to believe.
Amen."

When I'm Alone

When I'm alone
Is when I meet my Saviour face to face,
And praise Him for His wondrous love and grace.
He takes my cares and worries. In their place
He leaves a joy that nothing can erase.

When I'm alone
I think about the world He came to save,
About the loved ones that to me He gave,
And pray that somehow He will make me brave;
That I will always tell His power to save.

When I'm alone
He speaks to me with comfort if I'm sad;
I feel His love, and things can't be so bad.
I find a hope that I have never had,
And I go back to work with heart that's glad.

Your Will Be Done

If I should ask for sunshine
When flowers thirst for rain;
If I should wish for cloudy skies
When golden grows the grain;

If I should hope the frost to feel
Before the birds have flown;
Then grant not my desire, Lord.
Your perfect will make known.

If I should want my life set out
Upon a golden dish;
If I expect You, Lord, to grant
My every little wish;

If I would ask for heights of joy,
Yet shun life's deep despair;
Open my eyes, and help me see,
Of sorrow I need my share.

If I should ask for gifts, O Lord,
When I have much to give;
And if I want things I don't need,
When others barely live;

If I forget to be content
With what You've given me;
Then take away the things I have,
That I, Your will might see.

God's Love

If I would know You, Lord, I must admit
That I've done many wrongs, but mustn't quit.
I need a loving Friend to trust, confide in;
Someone Who can be strong, yet all-forgiving.

You're mine, dear Lord, and what a Friend You are.
You're by my side each minute, every hour.
You give me strength and confidence to face
Life's difficulties, not with fear, but grace.

You've loved me since the very Earth You formed.
You paid my debt of sin, left me unharmed.
To think upon this truth leaves my head bowed.
I know myself, and therefore, can't be proud.

Great Victory is Coming (Paradelle)

Hold your head high, great victory is coming.
Hold your head high, great victory is coming.
Look unto God, His help, it is nigh.
Look unto God, His help, it is nigh.
Great victory is nigh, His help, it is coming.
Look unto God, hold your head high.

Speak forth His Word with boldness and power.
Speak forth His Word with boldness and power.
This desperate hour, you will be heard.
This desperate hour, you will be heard.
You will be heard. Speak forth this hour,
With desperate power and boldness, His Word.

Fight in His strength, with courage and grace.
Fight in His strength, with courage and grace.
The challenge, to face the enemy boldly.
The challenge, to face the enemy boldly.
In His strength, the fight, and courage to face
The enemy boldly; challenge with grace.

The Word, it is victory; His coming is nigh.
With boldness and power, hold your head high.
This desperate hour boldly challenge with grace.
Fight in His strength, the enemy face.
Forth! Great His help! Look unto God.
Courage to speak, and you will be heard.

Treasures in Heaven (Rhyming Couplets)

Though I may own the finest clothes that money can afford,
They will not last beyond this life, though treasured and adored.

The car I drive may be the best that you can buy today.
To idolize a piece of steel that someday will decay?

The wealth and riches I've laid up, they are but fleeting things.
The day may come when all is lost. They'll fly, as if on wings.

This Earth is filled with moth and rust. Thieves everywhere abound.
Take note when storing earthly things, there is no safe place found.

The Bible tells us of a Home that will be ours someday.
Try sending treasures on ahead, where they will not decay.

The way to do this is to love the Lord with all your heart,
Then love your neighbour as yourself, the perfect place to start.

Take time to do God's work on Earth. Tell others of His love
Today, and long as you shall live. Thus, treasures build, above.

Inspired by Matthew 6:19-21

In His Image

The apple tree stands tall and strong
With leaves of green and blossoms white
Soon bees arrive to pollinate
And fruit begins to come in sight.

The apple grows to ripening
And falls upon the Autumn ground
To rot, or feed the worms and birds.
Before too long, no trace is found.

But deep inside the apple core
A seed, the image of that tree,
Waits patiently until the day
When finally, it will be free.

The seed lies dormant on the ground,
'Til sun and rain cause it to grow
Small roots take hold, green leaves appear,
A brand new tree begins to show.

And so it is with us and God.
We are the fruit, He is the tree,
We live and grow, we age and die,
Our spirits longing to be free.

The image of the Living God
Dwells deep within each human heart.
Our spirits, like the apple seed,
Set free, when earthly cares depart.

Good Versus Evil

Evil prowls like a stealthy cat.
He stalks his prey at night.
He'll kill and eat until he's fat,
Whatever comes in sight.

Goodness shines like the soft sunlight
That comforts one and all,
But should that evil cat get in,
His shadow casts a pall.

The sunlight never goes away,
But evil blocks its shining.
So too, in life, the choice is ours
To find that silver lining.

Let It Flow!

There's a well of living water
Springing up within your heart
Coming from the Lord Himself , so
Let it flow!

Bring God's Word to life within you
Read and meditate each day
'Til your heart is filled to bursting.
Let it flow!

There is healing, there is power
As the Spirit fills your heart
Drink until your thirst is quenched, then
Let it flow!

Many lonely people, thirsting
For His comfort and His love,
Are just waiting for someone to
Let it flow!

So why not take up the challenge?
Make this world a better place.
Spread the love of God around, just
Let it flow!

The Lord is Good (Rondeau)

The Lord is good, His mercy never ends.
His child He will protect, His love He sends.
The world is all around, temptations great,
But He will give His strength, if we just wait.
The Lord is good, so to His child He tends.

His Name is Jesus. He has made amends
For every sin, each time His child offends.
He gave His life, new life to re-create.
The Lord is good, His mercy never ends.
Each tear He dries, each broken heart He mends.

Compassion in His voice, His ear He lends.
His resurrection opened Heaven's gate.
Now we may dine with Him, we have a date.
The invitation, He to all, extends.
The Lord is good, His mercy never ends.

Who? (Quinzaine)

Beautiful warm summer day.
Who sends the sunshine
And the clouds?

Tall pines give shelter and shade.
Who makes them so strong
And mighty?

Green leaves rustle in the breeze.
Who tells the wind, "Blow,"
Or, "Be still"?

Bright flowers adorn hillsides.
Who clothes them with grace
And beauty?

Squirrels chatter endlessly.
Who provides them food
And comfort?

Birds and butterflies abound.
Who gives them their songs,
Bright colours?

Stream ripples peacefully by.
Who supplies water,
Maps the way?

Who?

The Prayer of Jabez (Pantoum)

Father, I desire Your extravagant blessing.
I pray that opportunities to serve You would arise.
I seek for Your guidance, that I need not be guessing
The path I should take. God, help me be wise.

I pray that opportunities to serve You would arise.
The way to choose good above evil, please show me.
The path I should take. God, help me be wise.
To never cause pain to those who would know me.

The way to choose good above evil, please show me.
That I might bring honour to Your Holy Name.
To never cause pain to those who would know me.
The life I am living, may it never bring shame.

That I might bring honour to Your Holy Name,
I seek for your guidance, that I need not be guessing.
The life I am living, may it never bring shame.
Father, I desire Your extravagant blessing.

(from I Chronicles 4:10)

Treadmill

We're on the treadmill of the world,
"Perform, perform, perform!"
With God it's not that way at all.
He holds us in His arms.

We rush about at work or play,
We, "shop until we drop."
We try to have it all today,
But God says, "This must stop."

"I love you just the way you are.
I cannot love you more
Your righteousness is filthy rags.
It's love I'm longing for."

Make the Choice

The city was blockaded. The famine was severe.
All food had long been eaten, and everything was dear.
Some Mothers ate their children! The King was seeing red!
He blamed it on the prophet, and vowed to have his head.

But there were four men (lepers, what did they have to lose?)
Who sat beside the city gate, and they began to muse,
"The famine's in the city. If we stay here we'll die.
So let's go to our captors. It's surely worth a try."

Away they went, discovering the enemy had fled,
Deserting all their worldly goods. The lepers soon were fed.
Then off to tell the King about the riches they had found.
Soon everyone was praising God. The wealth was passed around.

The moral of this story is, no matter who you are,
The lowliest, the weakest, an outcast from afar,
If you have faith and courage to listen to God's voice
He'll use you in a mighty way, but you must make the choice.

God is Speaking (Autoinversion)

God is speaking when we're dreaming,
When we're working, when we're scheming.
Never does His guidance cease,
Listen, and you'll find great peace.

There amid the voices screaming
Always quietly redeeming.
Listen, and you'll find great peace,
Joy and comfort, sweet release.

Care and worry soon will cease.
Listen, and you'll find great peace.
Down upon our hearts He's beaming
His great love, our souls redeeming.

Listen, and you'll find great peace.
Stop your striving, be released.
Put to rest the worries teeming
In your mind. His love is beaming,
Soul-redeeming.

Believe (Rhyming Poetry)

Believe that Jesus washed your sins away.
Believe, and start a brand new life today.
Believe that He will wipe away all tears.
Believe He'll take your cares and calm your fears.
Believe.

Believe, and find the strength to overcome
Those weaknesses that make your life so glum.
Believe, and He'll provide for all your needs.
He'll speak, and you will follow where He leads.
Believe.

Believe for opportunities to share.
Believe for needed boldness so you dare.
Believe in His salvation for your home,
That all your loved ones soon will cease to roam.
Believe.

Believe that by His stripes your body's whole.
Believe that He'll return what Satan stole.
Believe He'll prosper you as you obey,
Believe you'll dwell with Him on high some day.
Believe.

Dry Bones (Rhyming Couplets)

God led him to a valley deep, a valley filled with bones.
These bones were dry and dusty, as lifeless as the stones.

"Can dry bones live?" God asked him. "Lord, You would know," he said.
"Well, go and prophesy to them, and raise them from the dead."

The prophet spoke the Word of God. The bones began to shake.
They soon had joined together, a human form to make.

Then sinew came, and flesh and skin. God said, "Speak life to death."
He prophesied to wind and air, "Blow into them, O breath."

They rose and stood before the man, an army of great length,
All ready to do battle for the Lord with all their strength.

Now, God can do the same for us when we feel dead and dry.
Believe, receive the life He gives. It couldn't hurt to try.

He came to bring abundant life to all who would believe,
So leave behind your hopelessness. Just ask, and you'll receive.

Prayer for Governments (Deten Poetic Form)

I pray, O God, for governments,
For rulers of this worldly realm.
I ask that they would turn to Thee,
Prime Ministers and Presidents,
Lord, may they be not overwhelmed.

God, grant them wisdom to do right,
To stand against the Evil One.
I pray their goal would always be
To look to Thee for guiding light
And health and strength, the race to run.

I ask for Thy protection, Lord,
Upon Thy people, East and West.
I ask that Thou wouldst keep us free,
That we may live in one accord.
Thy will be done. Thou knowest best.

Circle of Life (Double Eintou)

"Birth" day.
Helpless infant
Soon grows to adulthood,
Marries, has children, works and plays.
All too soon old age calls,
Helpless again,
Death comes.

"New Birth"
Life from above.
"Old things have passed away,
Behold, all things have become new."
Spread the message to all.
Eternal home,
Heaven.

I Dreamed a Dream

I dreamed a dream, a wondrous dream.
The air was thick with fear, it seemed.
My son and I alone were found,
With enemy soldiers all around.

I could not move, I dared not run,
But suddenly, "Oh, no! My son!"
Away he went, but lo, behold!
No soldier tried to stop him cold.

At this strange turn, my boldness grew.
A new thought came, "Perhaps I too,
Could leave this place," and off I went
Like someone on a mission sent.

A sword of paper in my hand,
Great courage came. I took my stand.
I beat their chests with all my might,
The soldiers powerless to fight.

So we escaped. All fear was gone.
My paper sword the battle won.
One question lingers, "Please tell me,
The paper sword, what can it be?"

Love Sets Free

A little old man I met one day.
His back was bent, his hair was grey,
But on his face was a smile so gay.

He told me about his lovely wife
Whom he had caused much pain and strife,
Yet she had loved him as her life.

Then she had died, left him alone,
So for his crime he could not atone.
Such agony his heart had known.

I marvelled, as I asked him, "How,
If that is so, are you happy now?
He said, with a funny, little bow,

"Someday my dear wife I shall see.
Her prayers at last caught up with me.
I've been forgiven. God, too, loves me!"

Love (Decimeter)

Love, greatest of all gifts from God above.
Faith and hope cannot compare to love.
Fruit of the Spirit, Number One,
Encompasses the rest. None
Can conquer. Peaceful dove.
Patient, kind, won't shove.
Shines like the sun
Up above.
The Son,
"God."

Joy (Decimeter)

Jesus and you, with nothing in between.
Fruit of the Spirit that can be seen.
God gives us joy. Even when sad,
His loving care makes us glad.
Singing praises serene,
On His love we lean,
Righteousness-clad.
Lasting sçene,
No fad!
"Strength."

Peace (Decimeter)

Peace is like a bird singing in the rain,
Though storms rage on, we hear his refrain.
Gift from God, never understood.
Fruit of the Spirit, so good,
Serene, unruffled, plain,
Never turning vain,
Fruit that all should
Seek to gain.
We would
"Rest."

Patience (Decimeter)

Patience is learned by going through trial.
Helps us to endure for a long while.
Fruit of the Spirit, keeps us cool
Waiting, we won't act the fool.
Everything goes wrong? Smile.
Take what comes. Don't rile.
This is life's school.
One more mile
You rule.
"Calm."

Gentleness (Decimeter)

Gentleness is like a free, flowing breeze,
Whispering softly among the trees.
Fruit of the Spirit, always kind,
Flows from the heart, not the mind.
Opportunities seize,
To help out, with ease.
Never behind;
Seeks to please,
Not bind.
"Love."

Goodness (Decimeter)

Goodness is thinking, being, doing right,
Resisting evil, living in light.
Treating others with compassion.
Loving God with deep passion.
Fruit of the Spirit bright.
Joining in the fight
Temptations shun.
Pow'r and might,
War won.
"Pure."

Meekness (Decimeter)

Meekness is humility; thinking not
More highly of yourself than you ought.
Fruit of the Spirit with great worth -
Inheritance of the Earth.
No recognition sought.
Always gives great thought
To others' worth.
Can't be bought.
New birth,
"Heir."

Faithfulness (Decimeter)

Faithfulness is always remaining true,
Loyal, steady, counted on to do
Anything that needs to be done
Any weather, rain or sun.
Fruit of the Spirit, hue
Often called true blue.
Follow the Son,
Life made new,
Changed one,
"True."

Self Control (Decimeter)

Self control is always keeping your cool,
Will never make you look like a fool.
Fruit of the Spirit. Hesitate.
Think before you act, abate.
Found in the Golden Rule.
Count to ten, good tool.
Get rid of hate,
You'll be cool.
Don't wait.
"Love."

About the Author

I am a country girl at heart. I was raised on a farm, and loved the freedom of the country life and the peace of being so close to Nature. I also love God with all my heart and try to serve Him to the best of my ability in my daily living.

Our home was a happy one, with parents who were always there for us. We lived a carefree life until our Dad passed away when I was just 15, at which time I, as the eldest, had to grow up very quickly and take on much more responsibility. But we stayed on the farm, and were able to manage quite well, with the help of wonderful friends and neighbours.

I started writing poetry in my teens, and continued until my mid-twenties. Then, after thirtysome years of simply becoming too busy making a living and being involved in many other activities, I discovered I could still write. I joined a couple of poetry sites online, and for the past two years (2005-2007) I have written and reviewed thousands of poems. I have learned many styles of writing, and always enjoy trying something new.

Printed in the United States
95533LV00005B/4-6/A

9 781434 302434